37 Ways to Naturally Treat and Prevent Ovarian Cysts – The Facts and Myths About Herbal Healing

By Rosie Sanders

Introduction

We live in a time like no other. While we have technology beyond what our ancestors could have dreamed of, those advancements have come at a price. We have the ability to treat diseases more effectively, yet we seem to have more of them. Why is this?

Technology has given us many gifts. However, it also has given us some unwelcome side effects. In a world where chemicals are everywhere, we have now found that some of those chemicals imitate too well our own endocrine system. Hence, we have a rise in hormonally related problems.

This includes conditions like ovarian cysts, PCOS, undescended testes, early maturation, delayed menopause, and other hormonal disorders. We gain weight but can't lose it. We have trouble getting and staying pregnant. We experience pain and other symptoms no one should have to go through.

But what can be done about it? How can we reverse these negative effects?

Change has to happen on several levels. Government, to varying degrees, tries to regulate known dangerous chemical additives. But the ultimate responsibility falls to the individual.

We must educate ourselves as to which chemicals cause us harm. We must vote with our dollars how companies run their business. Most of all, we must take responsibility for our own well being.

No governmental controls, no corporations, and no insurance agencies are going to fix the health problems facing our culture today. We have to take action as individuals, and teach those we love to do the same.

Now, what does this have to do with ovarian cysts?

Traditional treatment plans for ovarian cysts consist of three major options: watchful waiting, hormonal treatment, and surgery. None of these options treat the true cause of the cysts. All of them are just like bandages, covering the symptoms yet failing to solve the problem.

Even if you take the most extreme route and have one of your ovaries removed, you could still get cysts on the other. If you have them both removed, you risk other diseases like breast cancer. Instead of going this route, why don't we pursue treating the cause?

Because it is not easy. It will take work, and you will have to experiment to find out which method works best for your body. In this book I've covered 37 different home remedies for treating ovarian cysts. Consider trying a few different ones. Take notes of what has worked for you and what has not. Through trial and error, you will be able to find the method which keeps your cysts under control.

The methods this book fully endorses are the following: cleansing, controlling blood sugar through diet, and exercise. These are generally the safest and most effective.

One more note: most of these methods have not been studied on pregnant women. In fact, very few scientific studies are done on herbal remedies in general. There is just not the funding to do so, since companies do not make a lot of money by promoting these methods.

So I propose we do our own study. If you can, please keep track of which methods you tried, how you used them, and what your results were. If enough of us participate, eventually we will come up with a more definitive answer.

You can contact me through my website: http://naturalremedyforovariancysts.com

What is a Cyst?

Before we go into treatment methods, a little understanding of the normal ovulation process is needed. In this chapter, we'll cover the normal egg release cycle, the different types of cysts, and their suspected causes.

First, let's go over what happens in the normal egg release cycle.

When a menstrual cycle starts, the brain sends a signal to the ovaries to start preparing eggs. This is called Luteinizing Hormone (LH). So sacs, called follicles, start growing around about a hundred of these eggs.

These sacs get bigger and start to move towards the edge of the ovaries. At that point, the ovaries produce estrogen. This tells the uterus to prepare itself for an egg, which it does through thickening the lining.

At around 14 days into this cycle, one of these follicles gets big enough to release an egg. This is called ovulation. The remaining sac is called the corpus luteum (yellow body). The corpus luteum then makes a hormone called progesterone. This causes the rest of the follicles to stop developing and be reabsorbed into the body.

Sometimes, this process goes wrong. The most common problem happens when there is not enough progesterone produced to tell the body to reabsorb the follicles. A few can remain and fill up with fluid, which creates the cyst.

Sometimes the corpus luteum does not dissolve. It can reclose and fill with fluid. This type of cyst is most common in pregnancy.

These types of cysts are known as functional cysts. They are the ones most likely to respond to natural treatments. If your body can be restored to optimal health, then they are easily restored.

About one in ten women have a condition known as polycystic ovarian syndrome (PCOS). This condition causes multiple cysts on the ovaries, as well as numerous other side effects. Many women have success controlling this condition through diet, exercise, and natural progesterone.

However, there are rarer forms o_
occasionally occur.   These types of cyst_
harder for the body to reabsorb. There have
been claims that some women have naturally
cured these types of cysts, but there is little
definitive proof. Most of these types of cysts
require surgery.

Dermoid cysts are rarely cancerous. These form
from your eggs. They can contain tissue such as
hair, skin, or teeth. They can become large, but
they generally grow slowly. They do need to be
removed surgically before they are too large, as
they will not dissolve on their own.

Endometriomas (chocolate cysts) form because
of endometriosis, a disease in which your
uterine cells grow outside your uterus. The
tissue can attach to your ovary and form a
growth which bleeds every month. If surgery is
done to remove these, they usually recur.
Generally, birth control pills are prescribed to
control this. There is some evidence that it can
be controlled through progesterone cream.

It is hard to diagnose endometiomas until they
are surgically removed. Many are, as they tend
to look like ovarian cancer on the ultrasound,
being solid. Endometriosis also raises your CA-

125 levels, which is another ovarian cancer danger sign.

There is some evidence that endometriosis can be controlled through eliminating excess estrogen and increasing progesterone levels. Estrogen makes the endometrial tissue grow, while progesterone helps it shrink. Thus, it makes sense to use these methods to control endometriosis and eliminate the cysts.

Cystandenomas develop from ovarian tissue and are usually filled with a liquid or mucus substance. These tend to become very large. Mucus filled cystadenomas grow larger than fluid filled. In rare cases, they may weigh up to 100 pounds.

You may also hear these terms thrown around in describing your cyst. A complex cyst is one that is composed of both solid and liquid material. A septated cyst is one that is divided by a wall. (Septum means wall.) Cysts with solid material are considered more likely to be cancerous. However, septation can occur with functional cysts.

Is It Cancer?

A lot of women ask if an ovarian cyst is a symptom of ovarian cancer. 95% of the time, an ovarian cyst does not indicate cancer. Fluid filled cysts are usually benign. In most cases, even solid cysts are benign. There are some tests your doctor can do to calculate the likelihood your cyst is cancerous.

You can get a blood test for CA-125. This indicator shows up in 90% of women who have ovarian cancer. You can also ask for a more recently developed test. The FDA approved the OVA1 test last year. It measures five different proteins that change because of ovarian cancer. It then combines the results into a score that indicates the likelihood your cyst is cancerous. Make sure you ask your doctor about it, as it is fairly new.

Your doctor will likely recommend surgery. However, if the results of the other tests are negative and you have no family history of cancer, you may want to try natural methods to shrink it first. This is very difficult decision to make.

Dangerous Symptoms

While most of the time your cyst will not cause any further problems, sometimes complications do develop. The most common are a burst cyst and ovarian torsion. Both of these require a visit to the hospital.

When a cyst becomes too large, it can burst. This can be dangerous because it may cause internal bleeding or infection. You'll need to go to the hospital for immediate treatment.

A burst cyst will cause severe abdominal or pelvic pain. Usually, when a cyst bursts, you have a stabbing pain, followed by a steady intense pain. The stabbing pain comes from the actual rupture; then, the fluid from the cyst will irritate the surrounding area. This may burn or feel achy. You need to see a doctor right away to start on antibiotics. The doctor will also want to monitor you to make sure you do not have internal bleeding.

If a cyst grows too large, this can cause ovarian torsion. The ovary twists on the fallopian tube, which causes severe, steady pain. If this happens, you need to see a doctor or go to an emergency room right away. This is a rare case, but one that can cause you to lose your ovary if you do not get treated right away. Treatment is surgery.

There may also be pain accompanied by fever or vomiting. In severe cases, you may have symptoms of shock, including rapid breathing, cold clammy hands, and lightheadedness. If this happens to you, you need to see a doctor right away.

Natural Treatments to Get Rid of Ovarian Cysts

The first methods I recommend using are the general tips. These increase overall health for everyone.   Combine these techniques with other recommended methods to see what works for your body.

Not all of these methods are going to work for you. Recalibrating your hormonal balance is a difficult task, and very hard to do without knowing where you start from. For example, many women have a low level of progesterone. So taking herbs which increase progesterone production may help. Some women produce too much testosterone. They need to take supplements which suppress androgen production. Many women with cysts have too much estrogen from the environment. They need to cut back on external estrogen, and cleanse the liver to help eliminate unwanted hormones.

To determine what works best, please keep track of what you try and how you feel afterwards. If one supplement makes your symptoms worse, then discontinue it and switch to another treatment.

You also will want to talk to your doctor before starting any supplements. Some can interact with other prescriptions you are taking, and others may make any underlying conditions you have worse.

Finally, do not try any supplements if you are pregnant or nursing. With each treatment, the known interactions are listed, but not every supplement has been extensively tested. Proceed at your own risk. Always talk to your doctor before you start any form of nontraditional medicine.

General Tips

Method 1 - Balance Blood Sugar

Safe For Everyone

Many women with ovarian cysts and PCOS have some level of insulin resistance. This means that the insulin your body produces Is not as effective at carrying glucose, or sugar, across cell walls.

Insulin resistance means your body has become less able to process blood sugar. Whenever we eat food, sugars from that food enter the blood stream. It then travels to all the cells in our body to provide energy.

Insulin is the key to getting the sugar from the blood into our cells. Sugar cannot pass by itself through the cell walls. It needs the insulin to unlock the doors so that it can get in, and leave the blood stream.

Your body wants to maintain the blood sugar levels at a certain level. When it gets too high, your pancreas secretes a lot more insulin.

When it's too low, then you will get a craving for carbohydrate rich or sugary foods.

When your body becomes insulin resistant, the cells in your body do not respond to insulin as well as they used to. So your blood sugar levels rise. The pancreas then releases a lot more insulin, which leads to a sudden drop in blood sugar, which causes you to crave more carbohydrates.

Eventually, the insulin becomes almost completely ineffective. This leads to excess blood sugar in the body, which becomes diabetes.

There are several ways you become insulin resistant. Some people are just that way genetically. For others, this develops over time if you eat a diet high in carbohydrates and refined sugars. Excess weight and a lack of physical activity can also contribute.

If this is left uncontrolled, you will likely develop diabetes. This also increases your risk of stroke, heart disease, and possibly even cancer.

You can control it through diet and exercise. The easiest way to do this through diet is to adopt a low glycemic index diet.

What that means is you want to primarily eat foods that are low in the glycemic index. This means staying away from high sugar foods, as well as processed foods. Replace them with whole grain breads, brown rice, and oatmeal. Eat plenty of fruits and vegetables. You also may want to eat five small meals a day to keep a steady blood sugar level.

Method 2: Avoid Meat and Dairy
Safe for Everyone

Most meat and dairy products are hormonally treated. Cows are injected with hormones to get them to produce more milk. Almost all farm animals are treated with antibiotics to keep them healthy in the situations they exist in.

These hormones can negatively affect your body, especially if you are prone to cysts. They can cause the development of cysts, or increase the chance that existing cyst will grow larger.

If you must eat meat and dairy, make sure you get only those that are hormone free. They are a little more expensive, but definitely worthwhile.

Make sure that you supplement your diet with calcium and other sources of protein if you use this method.

Method 3: Avoid Soy and Flax
Safe for Everyone

Most women do not know that soy and flax actually contains phytoestrogen. This is a plant based form of estrogen that closely mimics your body's own estrogen.

Phytoestrogen is able to bond to your estrogen receptors. This can wreck havoc with your endocrine system, especially if you are sensitive to it. Having too much estrogen also increases your risk of breast and ovarian cancer.

This form of estrogen is thought to be somewhat safer than birth control pills. However, it is not what is needed to get rid of ovarian cysts. It can actually make a cyst increase in size, especially endometrial cysts.

It's wise to avoid these until your cyst has dissolved. Other foods that contain estrogen include:

- French bean
- Date palm
- Dates
- Garlic
- Pomegranate
- Apple
- Soyabean
- Chick pea
- Cherry
- Alfalfa
- Soya sprouts
- Green beans
- Red beans
- Split peas
- Flaxseed
- Raspberry
- Carrot and squash (both have beta carotene).

These foods all have small amounts of estrogen, so they should be safe in small portions. Try to limit yourself to one serving of any one of these a day if you cannot avoid them. These are very healthy foods and should be reintroduced into your diet after your cyst has dissolved.

Method 4 – Wash Fresh Fruits and Vegetables in Vinegar
Safe for Everyone

This method may sound odd, but it can help. Most fresh fruits and vegetables are treated with pesticides. Pesticides contain xenoestrogens, which can aggravate cysts. Washing them in one part vinegar and one part water is more effective than just washing in water to remove these pesticides.

An easy change to make, and one that can vastly improve your health!

Method 5 – Drink Plenty of Water

Safe for Everyone

Did you know that most people walk around in a constant state of dehydration? This puts undue stress on the body, and causes a number of ailments.

Most people need to drink 8-12 glasses of water a day. Each glass should be 8 oz, or 1 cup. Most of us drink 4-5 glasses a day.

To remedy this, try increasing your water consumption by one glass a day until you are drinking at least 8 cups. You will have to go to the bathroom a lot at first. This is because your body uses the fresh new water to flush out the old water that it was storing.

After a few days, this should level out.

One thing to watch: most tap waters contain trace amounts of a large number of dangerous chemicals, prescription drugs, and other additives. The majority of tap water is treated with chlorine, which has been linked to the development of cancer.

You will want to either invest in good bottled water or a good water filter. Reverse osmosis filtration is considered to be the most effective.

Method 6 – Exercise
Safe with doctor's permission

Exercise is an extremely effective way to restore health. Just working out for thirty minutes a day, five days a week, can have enormous effects on strength, endurance, energy, and mood. Working until you sweat allows your body a chance to excrete out impurities through the skin.

Most of us have at least a few pounds to lose. Combined with a nutritious diet, exercise can help you obtain a healthy weight. This can help your body better regulate blood sugar and insulin levels, which in turn, help prevent cyst formation.

Many women believe they don't have time for exercise. That is completely understandable! However, unless you make some time for exercise, you will find yourself in poor health and unable to do what you want.

Exercise does not have to be hard to be effective. The key is to find a physical activity that you enjoy. Swimming, yoga, gardening, dancing, and walking are all ideas.

You want to combine activities such as the examples with weight training. Weight training has been shown to improve bone density. It also builds up muscle, which burns calories even when you are sitting still. It's one way to maximize your time spent working out.

Method 7 - Detoxification

Safety depends on method used – talk to your doctor before trying

Since we are exposed to so many toxins on a daily basis, it is a good idea to practice detoxification. Detoxifying the body allows chemicals to be processed quicker. It also clears out extra estrogenic compounds, which contribute to the formation of cysts.

A liver detoxification is especially important. This cleanses the liver of gallstones, which clears it to better process extra hormones.

Before you do a liver cleanse, you will want to do a parasite cleanse. Otherwise, the liver cleanse could make you very sick.

Make sure you find a method that is highly reviewed and recommended. There are several kits available at health food stores and online.

You may also want to do a colon cleanse. This will eliminate accumulated fat and waste material from the colon, and allow it to operate more efficiently. After these cleanses are done, make sure to continue a healthy diet to keep your body operating at its full potential.

Pain Relief

Method 8: Hydrotherapy

Caution with Pregnancy or Vaginal/Uterine Infections

Hydrotherapy just means you treat with water. It is believed to help increase circulation, relieve pain, and help facilitate reabsorbtion of a cyst.

The ideal treatment is to do a contrast sitz bath. For this, you need two containers, one filled with cool water and the other filled with hot water. Remember that hot water will feel cooler to your hands than the rest of your body, so make sure you test the temperature on your wrist.

You can get a basin that fits over the toilet or just use the tub. The toilet one can be more convenient to use.

Start by sitting in the warm water for three to four minutes, then soak in the cold water for thirty seconds. Repeat five times.

Method 9: Castor Oil Pack

Not Safe For: Pregnant, Nursing, and Menstruating Women – MAY INDUCE LABOR! Do not use over broken skin, or if you have a tumor, uterine growth, or an ulcer.

The castor oil pack works by relaxing smooth muscle, including the uterus and blood vessels. It relieves pain, increases circulation, and promotes healing and elimination of toxins in the organs under the skin. They work very well on ovarian cysts, fibroids, and even constipation.

If you don't want to make a pack, you can just put castor oil onto your skin. Cover it with a cloth you won't mind discarding, and put your hot water bottle or heating pad on top of that. You may want to have an old towel or plastic wrap underneath of you to catch any drips.

Do this three to five times a week until you see improvement.

Materials needed:

- Old towel or flannel cloth
- Plastic sheet (trash bag fine)
- Castor oil

- <u>Heating pad</u>
- Second old towel
- Large plastic bag

Take a terry cloth towel and fold over on itself four times. It should be about four inches wide. You can also use a flannel cloth if you don't want to ruin a towel.

Saturate the cloth with the castor oil. It doesn't need to be dripping, but it should be soaked through.

Place the plastic and old towel over the oil pack to prevent staining.

Lie down, putting the soaked surface against your skin with the plastic on the outside.

Wrap the cloth and plastic around you.

Place a heating pad set on low heat or hot water bottle over your abdomen on top of the castor oil pack.

Rest for 30 to 60 minutes.

Afterwards, try to stay relaxed and still. It's best to do these before bedtime.

You can reuse the oil pack by folding it up and putting it in a plastic bag, then storing it in the refrigerator. It can be used until it starts to smell stale. If you get it out of the fridge to use again, let it warm up to room temperature for 1 to 3 hours before using.

Method 10: Willow Bark

Not safe for pregnancy; breastfeeding; children. May interact with anticoagulants, beta blockers, diuretics, NSAIDS, Methotrexate and phenytoin.

Willow bark has been used since the time of Hippocrates (400 BC). It reduces pain and inflammation through salicin, a chemical similar than aspirin.

These results have been confirmed through several scientific studies. White willow is a little slower to start working to bring pain relief, but its effects seem to last longer.

Studies have also shown that willow bark may have antioxidant, antiseptic, immune boosting, and fever reducing properties. Other studies have shown it to be as effective as aspirin at a much lower dose.

Studies also suggest that it may be less likely to cause stomach upset than other pain relievers, but this proof is not yet definitive.

The recommended dosing size is 60-240 mg in a capsule or liquid. Start with a low dose and increase until effective. Don't go over 240 mg without talking to a doctor.

There seem to be only mild side effects with this herb. Some potential side effects are stomach upset, ulcers and bleeding. Overdose may cause a skin rash, stomach inflammation, tinnitus (ringing in the ears), kidney inflammation, nausea and vomiting.

Nutritional Supplements

Method 11 – Vitamin C

Safe for everyone in recommended doses.

Vitamin C can also be helpful in reducing inflammation. This can help reduce cysts by lessening an allergic reaction to irritants. With endometriosis, this can significantly help.

It's best to get your vitamins from food sources. The recommended daily amount for women is 75 mg. Good sources of vitamin C include oranges, guava, red pepper, kiwi, green pepper, grapefruit juice, tomato juice, and strawberries.

If you get a supplement, it's usually recommended to get one with bioflavonoids. These help balance estrogen levels, which can keep cysts from growing. You can find naturally occurring bioflavonoids in garlic, cherries, onions, grape seeds, peppers, citrus, and currants.

Method 12 – Zinc

Safe in recommended dosages

Zinc is known as a booster for the immune system, but it also may help the reproductive system. It is necessary for normal egg development. It also can protect against free radical damage. These free radicals can cause cell damage. Zinc helps to keep them under control

This on its own is probably not enough to clear away ovarian cysts. However, it is a good supplement to add in. You can find it in oysters, beef shanks, crab, pork, and fortified cereal.

Method 13 – B Vitamin Complex

Talk to your doctor before taking if pregnant or breastfeeding; have allergies, anemia, or are receiving dialysis. May interact with hydantoins or levodopa.

The B-complex vitamins are used by your liver to convert estrogen into weaker forms. This can help your cyst to clear up faster.

Good food sources of this vitamin complex include whole and enriched grains, potato, dairy products, sunflower seeds, pork and dried beans.

Method 14 – Vitamin E

Safe for pregnancy if in proper dosage. Not for those with vascular diseases or cancer.
May interact with warfarin (Coumadin)

Vitamin E is an antioxidant which is often touted for its ability to clear the blood of free radicals. Since many studies have shown that a diet which contains antioxidant-rich vegetables and fruits are associated with lower disease rates, people have taken that to mean supplements give you the same results.

However, studies have shown this is not the case. These supplements are usually synthetic. Even when they are extracted from food, they lose some of their effectiveness. With Vitamin E specifically, the supplement form had little to no effect in improving a number of conditions.

That being said, a lack of vitamin E causes problems in the ovaries first in rat studies. They lost the ability to secrete enough progesterone, which may contribute to the formation of cysts.

There are several good sources to get vitamin E in your food. Almonds, sunflower seeds, hazelnuts, peanut butter, peanuts, corn oil, spinach, and broccoli all contain vitamin E.   It can be a good idea to consume more of these foods in order to increase your vitamin E level in case a deficiency is contributing to cyst formation.

Herbal Remedies

Method 15: Dandelion

Not Safe For: Those allergic to daisies, those who have gallbladder ailments. May irritate stomach. Diabetics should monitor their blood sugar when using.

Dandelion leaves are known as a diuretic. It also works to purify the liver, where a number of toxins are stored. This helps your body to better process toxins, which can reduce the occurrence of ovarian cysts.

You can find dandelion almost anywhere. However, if you pick it outdoors, make sure it has not been sprayed with any lawn chemicals.

The dandelion leaves are bitter, but they are less so if you pick them before the plant blooms or after a frost. For this treatment, you will want to use the leaves to make tea. They will need to be dried before doing so. You can either air dry them, or dry them in low heat overnight in the oven.

To make tea, mix 2 tsp dried dandelion leaf with 8 oz of water. You can add in a sweetener before drinking to cut the bitterness. Try a cup a day.

Side effects are limited. Occasionally, some people are allergic. There are also a few reports of upset stomach and diarrhea.

Method 16: False Unicorn Root
Potentially Unsafe for Pregnancy

False Unicorn is a traditional Native North American remedy. It was used as a uterine tonic in the US from 1916 to 1947. Today, it is used for conditions affecting the uterus and the ovaries.

It seems to have a normalizing effect on the reproductive system, encouraging a regular menstrual cycle. It also encourages the ovaries to release progesterone at the right time of the month. It is also used to treat endometriosis, so this herb may be useful for those with endometriomas.

False unicorn root contains steroidal saponins, which are the active ingredient. To take it, put 1 to 2 teaspoonfuls in a cup of water, boil, then simmer gently for 10 to 15 minutes. It should be drunk three times a day. If you purchase the tincture, take 2-4 ml of the tincture three times a day.

Method 17: Black Cohosh
Not Safe for Liver problems, breast cancer, pregnancy (may cause miscarriage).

This herb is commonly recommended for ovarian cysts. However, it may not be a good treatment.

Black cohosh has been shown to have estrogen like effects. It is commonally used by women in menopause to relieve hot flashes and other symptoms. This herb has had several studies done with mixed results.

Most showed slight increases in LH levels, but they were not statistically significant. It did relieve the frequency and intensity of hot flashes. There was also some improvement in mood.

However, black cohosh is believed to cause liver failure in at least one case. It is believed that it may damage the liver over time, even at small doses. This has not been studied conclusively.

This method is included since it is commonly recommended, but this book does not recommend it. There are safer and more effective ways to treat ovarian cysts.

Side effects include stomach pain, headache and rash. If you start to have liver trouble, such as abdominal pain, dark urine, or jaundice, stop taking this and immediately go to the doctor.

Method 18: Echinacea

Not safe for those allergic to ragweed, marigolds, or similar plants; autoimmune disorders; HIV; white blood cell or collagen disorders; tuberculosis. Those taking immune suppressing drugs or steroids should not take. Has not been studied on babies and small children, so may want to avoid if pregnant or nursing.

Echinacea is well known as an immune system booster.   This may help ovarian cysts by increasing white blood cell production, which then helps the body eliminate abnormal cells. It also may help restore hormonal balance within the body.

However, this has not been studied on women with ovarian cysts. So there is no conclusive evidence whether it helps or not.

If you do use Echinacea, make sure you do not take it for longer than eight weeks at a time. This herb becomes less effective the longer it is taken. It can cause liver damage if taken for too long, especially if you are taking steroidal medications.

There were other, less common side effects also reported with Echinacea usage:

- Stomach upset
- Nausea or dizziness
- Increase in asthma
- Allergic reactions, including rashes and swelling
- Fever

If you want to try this remedy, take the supplement according to package direction. Different providers will have differing concentrations. Most of the time, 900 mg of the dried powdered root or 6 to 9 ml of the pressed juice are effective.

Method 19 – Milk Thistle

Not Safe For Pregnancy; Breast Feeding; Ragweed allergies; liver, gall bladder or kidney disease. May interact with other medications and alcohol.

Milk thistle is one traditional medicine that has been used for thousands of years. It is native to the Mediterranean, and has been used especially for liver problems. It is claimed to help lower cholesterol levels and reduce insulin resistance. Some people have also claimed it reduces the growth of hormonally related cancer cells, such as breast, cervical and prostrate cancer.

Milk thistle is believed to help ovarian cysts by balancing a woman's estrogen and progesterone levels. It also is believed to work on the liver because it supports the liver's detoxification process. During this time, abnormal cells are destroyed and extra hormones, including excess estrogen, are excreted. It also is believed to stabilize ovarian function.

However, milk thistle is an estrogenic herb. It contains phyto estrogens. If you decide to take it to cleanse the liver, only take it for up to two weeks. Longer than that may actually worsen ovarian cysts.

It is recommended to take 280-450 mg a day In two doses. Other side effects may include stomach upset and diarrhea. It also may lower blood sugar levels, which can negatively affect those with diabetes.

Method 20 – Vitex (Chasteberry)
Not safe for pregnancy; nursing;
May Interact with Dopamine agonists
(bromocriptine, levodopa). Also makes hormonal
birth control less effective.

Vitex, or Chasteberry extract, is often used to treat hormonal imbalances in women. It stimulates and normalizes pituitary gland functions and progesterone levels. It can help balance the ratio of hormones required for normal ovulation and dissolution of the follicle.

Chasteberry is believed to be especially effective for women with PCOS, as it increases your progesterone levels. It is suspected that women with PCOS do not produce enough progesterone, which causes multiple cysts. Increasing the levels of this hormone can significantly alleviate these cysts.

This is also effective for endometriosis. Progesterone helps to shrink the endometrial tissue. This includes both the uterine lining and the tissue on the ovaries which cause the cysts.

Possible side effects include acne, cramping, diarrhea, hair loss, head ache, increased menstrual flow, stomach pain, or tiredness. Some people may be allergic.

Dosing depends on the brand. However, 700 mg is a commonly recommended dose.

This appears to be one of the more effective herbal treatments, but there has not been conclusive scientific studies on the subject yet.

Method 21 – Red Clover

Not safe for pregnancy, nursing. Also not safe for those with heart, liver, or kidney conditions; history of breast, uterine, cervical, or ovarian cancer; blood clotting conditions; migraines; seizures. May interact with hormonal medications.

This is another herb often recommended for ovarian cysts. It was believed that it helped to thin the walls of the cysts, allowing them to be easily absorbed. It also is supposed to act as a blood cleanser.

Side effects include nausea or upset stomach. Some people may also experience allergic reactions.

However, this is another phytoestrogen. It will make you feel better for about two weeks, then you feel much worse. Try other methods first. If you do use it, use only for short periods of time.

Method 22 – Wild Yam

Unknown effect on unborn and nursing infants; Considered safe otherwise.

Wild yam has long been believed to help women with ovarian problems. It is believed to help increase progesterone levels, and help balance hormones overall.

It may help regulate menstrual periods. It also is used to treat menstrual disorders, infertility, fibrocystic breasts, and ovarian cysts. Wild yam is often recommended as a substitute for birth control in treating ovarian cysts.

Usually, this is found in cream form. You can find it being sold as progesterone cream. Caution is needed when purchasing. Some supplements have been found to be contaminated with toxic metals or other substances. Make sure you purchase from a reliable source.

This is one of the most effective ways of clearing up ovarian cysts, especially in women with PCOS and endometriosis. The progesterone signals to a follicular cyst that it is time to dissolve. It also inhibits the growth of endometrial tissue, and even causes it to shrink. This helps chocolate cysts to dissolve.

You generally want to follow the dosing instruction on the package. Creams are believed to be more effective when applied in the pelvic area. Apply from days 14-26 of the menstrual cycle. Try to vary the point of application to avoid skin irritation. You want to use ¼ to ½ tsp a day on the thighs and the hips.

You can also take it in pill form, or as a liquid. Just don't take more than one form at the same time. This can result in an overdose.

Some people are allergic, but this is rare. There are seldom any reported side effects with this supplement. It also has not been reported to interact with medications.

Method 23 – Bee Pollen
May Cause Allergic Reactions

Bee pollen is growing in popularity as a health supplement. It is being promoted as a means to boost the immune system, increase energy, and promote overall health.

But what is bee pollen? It simply the pollen bees have collected from flowers and compressed into a small globe. It's taken back to the hive and used as a food source for young bees. It is collected by putting a wire mesh over the entrance to the hive. Some of the pollen is brushed off the bee as it passes by, and it is then collected.

You can find bee pollen raw or processed to remove impurities. There is no evidence that processing reduces allergic reactions though.

Some claims of bee pollen benefits include lowering bad cholesterol, protecting against breast tumors, increasing the immune system, stimulating the ovaries and protecting the eggs, and improving energy. It is also believed to slow the effects of aging.

It has been shown to be a powerful antioxidant, which can help remove toxins from the body. This can help the body to reduce cysts and improve overall health.

This is not for anyone with severe bee allergies or allergies to plants. If you take it and start to have an allergic reaction, go to the hospital right away. It can cause anaphylactic shock in those allergic.

This alone will not cure ovarian cysts, but it can help you restore overall health.

Method 24 – Bach Flower Remedies

May cause allergic reactions

Bach Flower Remedies are a very untraditional treatment method. Dr. Edward Bach practiced medicine in the early 1900s. After working as a researcher in immunology, he became interested in homeopathy. He then turned to alternative therapies and developed the Bach Flower Remedies.

He believed that symptoms of an illness are an effect of disharmony between the body and the mind. The symptoms are manifested from a negative emotional state. He developed 38 different natural remedies in all.

Robert McDowell recommends the following remedies in addition to herbal medicine for treating ovarian cysts:

- Walnut
- Wild oats
- Impatiens
- Red chestnut
- Honeysuckle

To take these, mix a 1 ounce bottle of fresh filtered water with 2 drops of each remedy. Also add 1 tsp of apple cider vinegar as a preservative. From this bottle, take four drops four times a day.

There are no known side effects for this method. It could cause allergic reactions in those allergic to nuts.

Method 25 – Evening Primrose Oil

Unsafe for epilepsy, schizophrenia, blood-clotting disorders, or before surgery. Unknown effects on unborn and young infants. May interact with medications for schizophrenia and anti-depressents.

Evening Primrose is a yellow flowering plant which blooms in the evening. It has been used since the 1930s for eczema. It also is used for inflammation and menstrual pain. It also may have some effect on endometriosis.

The oil from this plant contains gammalinolenic acid (GLA), which is the primary active substance in the plant. This is believed to help fight inflammation.

For best results, it should be taken orally. There are pills available, as well as tinctures. Make sure to buy from a reputable manufacturer, as some samples have been tainted with heavy metals. Don't use more than one variety of this oil at the same time. This oil has a calming effect, so it may help relieve stress, which can reduce ovarian cyst size.

There are very few side effects with this supplement. Some people may have an allergic reaction. Also, this may cause people on certain medications to have a seizure.

Method 26 – Dong Quai

Not safe for pregnancy and breastfeeding
Avoid using if have bleeding or clotting disorder, or if taking medicines to treat these. May cause allergic reactions. Avoid taking with warfarin, aspirin, NSAIDS (ibuprofen), ardeparin, dalteparin, danaparoid, enoxaparin, and heparin. May interact with blood thinners, hormonal medications, and St. John's Wort.

Dong Quai, also known as Chinese Angelica, is an herb commonly prescribed for many gynecological disorders. It is also used to relax muscles, decrease blood pressure, and treat constipation. This is one of the most popular herbs used in Chinese medication, even more popular than ginseng.

One US study found that it had little effect at alleviating postmenopausal symptoms. It has not been fully studied, so we don't completely know what the benefits of this herb are.

It is believed to help increase blood circulation in the uterine area and relieve pelvic congestion. It also is believed to help speed up tissue repair. Besides all this, it is also purported to stabilize blood sugar levels and increase immune system functionality.

One source did say it contains estrogen, but supporting information to back up this claim could not be found. Researchers are not yet sure on that issue.[1]

Usually, 3-4 grams of the herb a day is prescribed. You can take it as capsules, tablets, or as a herbal tincture. Do not use more than one form at once.

---

[1] http://www.umm.edu/altmed/articles/dong-quai-000238.htm

The biggest side effect of dong quai is an increased sensitivity to sunlight. It can make you much more sensitive to sun, so avoid excessive exposure. It also interacts with a variety of prescriptions and other herbs.

Method 27 – Licorice Herb

Not safe for pregnancy; breast feeding; heart disease; high blood pressure. May interact with diuretics, corticosteroids, and other medicines that reduce potassium levels in body.

Licorice should not be confused with the popular candy Twizzlers. While these are popularly known as licorice, Twizzlers contain Anise flavoring, a very safe food product. Licorice can actually be very dangerous in large amounts, as it reduces potassium levels in the blood. Too much can be fatal.

However, this herb can be an effective way to kill germs and viruses. It also has mild estrogenic properties, which lead to it commonly being prescribed for ovarian cysts.

However, like every other estrogenic herb, this may actually worsen cysts. Combined with the potential risk of heart problems, it is not recommended for treating ovarian cysts.

Method 28- Damiana

Not safe for pregnancy, breastfeeding, Alzheimer's disease, Parkinson's disease, schizophrenia, or a history of breast cancer. May interact with medicines that control blood sugar levels or treat diabetes.

Ancient Mayans used damiana as an aphrodisiac. It is still used today for its stimulatory action, especially in Mexico. When used, it produces a mild high.

There is no evidence that it helps ovarian cysts, although it is occasionally prescribed. It does help with depression. There are very few studies done on this herb, but it appears to increase your testosterone level. Thus, women with PCOS may want to avoid taking this.

You can take the leaves of this plant as an herbal tea, as a pill, or as a tincture. Make sure you only use one type of dosage at a time.

The only side effect is a sensation of euphoria when taken at a higher dose. It also has a mild laxative effect. Rarely, some people may be allergic.

Method 29 – Gotu Kola
Not safe for pregnancy; breastfeeding; diabetes; high cholesterol

This herb is popular in Sri Lanka. After observing elephants feeding extensively on the plant, they believed that gotu kola helped promote longevity. It has been used to treat people affected with mental problems, high blood pressure, excesses, rheumatism, fever, ulcers, and leprosy. It has now become a staple in Ayurvedic medicine. Most recently, it has been used as an aphrodisiac.

It is believed to help regulate hormonal levels, which is why it is often prescribed to treat ovarian cysts. However, no studies have been done to see if it is effective.

To take this medication, follow the manufacturer's directions. It is available in capsules and tinctures.

There are rarely any side effects. Sometimes it may cause nausea if the dosage is too high.

Method 30 – Ovarian Cyst Tincture (Burdock, vitex, raspberry, and motherwort)
Safety and Interactions Unknown

Occasionally, you will come across different remedies purporting to cure ovarian cysts. Some of these may be effective. Other ones may be dangerous. Most try to balance your hormones and cleanse the body.

Here is one you may want to try.

Ovarian Cyst Tincture

1 teaspoon each tinctures of burdock root, vitex berries, red raspberry leaves and motherwort leaves

½ teaspoon each tinctures of prickly ash bark and ginger rhizome

Combine ingredients. Take half a dropperful 2 or 3 times a day.

The effectiveness of these are seldom studied scientifically. However, you can find testimonials that they work[2].

Method 31 - Cinnamon

Some allergic reactions

Cinnamon is used in Chinese medicine to treat ovarian cysts. This is because it is believed to form from abdominal fluid accumulation. Cinnamon is believed to be invigorating, which balances out the cold static cyst.

2

http://www.mothernature.com/Library/Bookshelf/Books/15/67.cfm

There is no evidence cinnamon that helps ovarian cysts directly. However, it does help regulate blood sugar. In women with PCOS, excess blood sugar and a little resistance can contribute to the formation of ovarian cysts. Cinnamon, in small doses, can help control this.

You can include cinnamon in your diet by adding it to oatmeal in the morning, or putting a stick in your tea.

Method 32 – Peony

Potential unsafe while pregnant or breastfeeding. May cause uterine contractions. May slow blood clotting. Interacts with anticoagulant drugs and phenytoin (Dilantin)

Peony is another popular Chinese herbal remedy. It is used to treat PCOS, as well as other conditions. Generally, the root is used.

It is believed that peony blocks muscle cramps, prevents blood clotting, and acts as an antioxidant.

It does appear to be safe for short term use. However, it can cause stomach upset and cause a rash in people with sensitive skin.

If you choose to try this herb, use the recommended manufacturer's dosage.

Method 33 – Saw Palmetto

Not safe for pregnancy and breastfeeding; liver disease; heart disease; heart rhythm disorder, stomach ulcers; asthma; blood clotting disorder; Crohn's disease. May interact with blood thinners; flutamide; garlic or gingko biloba; birth control; iron supplements; hormone replacement therapy; NSAIDs; blood clot medication.

Saw palmetto is a palm-like plant which grows in the southeast United States. From the berries, a capsule form of saw palmetto is made. This is used to treat an enlarged prostrate, bladder irritation, and PCOS.

This herb suppresses androgen, or testosterone, production. This can lessen some of the side effects experienced by women with this disorder.

There are currently studies looking at its effects on cancer cells, but no studies have been done regarding its effectiveness at controlling PCOS.

It is usually well tolerated, but may cause mild stomach discomfort. Some people can be allergic – go to the hospital if you have an allergic reaction.

Method 34 – Apple Cider Vinegar
Safety Unknown

One folk remedy mentioned many times on the Internet is Apple Cider Vinegar. With all the testimonials, you would think there is nothing that this liquid can't do!

Vinegar is known as a preservative and a cleansing agent. It is believed to help fight infections, lubricate the joints, and help the body maintain a proper acid-alkaline balance.

The last benefit is perhaps the most useful in treating ovarian cysts. Since it is believed that women with ovarian cysts have a slightly acidic blood level, then apple cider vinegar can help restore the body to a neutral pH.

The recommended dosage is 1 tsp in a pint of water. Some women claimed that their cysts cleared up within three weeks, but this is not scientifically verified.

Side effects stated were a feeling of heaviness in the legs and womb, as well as mild cramping in the uterus.

Method 35- Beet tonic
Safety Unknown; May cause allergic reactions

Another remedy popular on the Internet is a concoction of beet, aloe vera juice, and molasses[3]. Numerous women have claimed that it is effective, although one experienced no improvement.

There were several recipes involving beets on the site, but here is the most popular:

- o 2-3 slices of beets, mashed.
- o 1 Tablespoon Aloe vera gel or liquid. Can substitute carrot juice.

---

[3]

http://www.earthclinic.com/CURES/ovarian_cysts.html

- 1 tablespoon Molasses

Drink once or twice a day for up to six weeks.

There is no scientific evidence for or against this method.

Method 36 – Potassium Iodide (SSKI)[4]

Unsafe for pregnancy; breastfeeding; Addison's disease; cystic fibrosis; tuberculosis. May cause allergic reactions. Talk to your doctor is you have hyperkalemia, thyroid problems, goiter, or kidney problems.

Potassium iodide, or SSKI, is most well known as a way to protect the thyroid gland from radiation injury. It's used in accidental exposure to radiation, as well as before administration of radioactive iodide.

It is also used to treat and prevent infection from water born bacteria. More recently, it has been used to treat breast and ovarian cysts.

---

[4] http://www.tahomaclinic.com/iodide.shtml

Iodine can help to metabolize bad forms of estrogen into good forms. Estrogen can be found as estrone (a carcinogenic human estrogen), and 16-alpha-hydroxyestrone (a more dangerous form). This can help the cysts to dissolve. It also has been shown to reduce inflammation and kill bacteria.

Iodine can stain, so be careful to not get in on your clothing. Some people may be allergic. This usually shows up as a red bumpy skin rash. To see if you are allergic, test on your skin first before taking internally.

Too much iodine for too long can lessen thyroid function. You only want to use this for a few days or a week or two. If you use it for longer, make sure to monitor your thyroid function.

Method 37- Chinese Bitters

Not safe for nursing or pregnancy; don't use with high blood pressure; talk to doctor about possible interactions.

Chinese bitters are a combination of two traditional Chinese herbs – gentian root (Long Dan Cao) and bupleurum root (Chai Hu). They are believed to help cleanse the liver, circulate the qi of the chest, and to clear heat and damp from the liver and gallbladder.

These two herbs taken once a day on an empty stomach have been effective for some women at reducing endometriosis and cyst size. However, scientific evidence is still lacking.

If you try this remedy, look for tablets, capsules or water extraction. Avoid tinctures or alcohol extractions, as this can limit the herbs' effectiveness at cleansing the liver.

Some side effects can include headache, nausea and vomiting if overdosed. Call the doctor if this happens. Some people may also experience sedation and drowsiness. A few may be allergic.

References:

Agency for Healthcare Research and Quality. *Milk Thistle: Effects on Liver Disease and Cirrhosis and Clinical Adverse Effects*. Evidence Report/Technology Assessment no. 21. Rockville, MD: Agency for Healthcare Research and Quality; 2000. AHRQ publication no. 01-E025.

Bee Pollen. Review of Natural Products. factsandcomparisons4.0 [online]. 2004. Available from Wolters Kluwer Health, Inc. Accessed April 16, 2007.

Bisset NG. *Herbal Drugs and Phytopharmaceuticals*. Stuttgart, Germany: Medpharm Scientific Publishers; 2004:534-536.

Black cohosh (*Cimicifuga racemosa [L.] Nutt.*). Natural Standard Database Web site. Accessed at www.naturalstandard.com on April 10, 2009.

Black cohosh root. In: Blumenthal M, Goldberg A, Brinckman J, eds. *Herbal Medicine: Expanded Commission E Monographs*.

Newton, MA: Lippincott Williams & Wilkins; 2000:22–26.

Black cohosh. Natural Medicines Comprehensive Database Web site. Accessed at www.naturaldatabase.com on April 10, 2009.

Borrelli F, Ernst E. Black cohosh for menopausal symptoms: a systematic review of its efficacy. *Pharmacological Research*. 2008;58(1):8–14.

Borrelli F, Ernst E. Black cohosh for menopausal symptoms: a systematic review of its safety. *American Journal of Obstetrics & Gynecology*. 2008;199(5):455–466.

Bupleurum. Review of Natural Products. Facts & Comparisons 4.0. http: / / online.factsandcomparisons.com / MonoDisp.aspx?monoID=fandc - rnp - 5062&quick=bupleurum&search=bupleuru m&disease=. October 2009

Chaste tree fruit. In: Blumenthal M, Goldberg A, Brinckman J, eds. *Herbal Medicine: Expanded Commission E Monographs*. Newton, MA: Lippincott Williams & Wilkins; 2000:62–64.

Chasteberry (*Vitex agnus castus*). In: Coates P,
Blackman M, Cragg G, et al., eds.
Encyclopedia of Dietary Supplements. New
York, NY: Marcel Dekker; 2005:95–103.

Chasteberry (*Vitex agnus-castus*). Natural Standard
Database Web site. Accessed at
www.naturalstandard.com on August 14,
2009.

Chasteberry. Natural Medicines Comprehensive
Database Web site. Accessed at
www.naturaldatabase.com on August 14,
2009.

Chrubasik S. Pain therapy using herbal medicines
[abstract]. *Gynakologe*. 2000;33(1):59-64.

Circosta C, Pasquale RD, Palumbo DR, Samperi S,
Occhiuto F. Estrogenic activity of
standardized extract of Angelica sinensis.
Phytother Res. 2006;20(8):665-9.

Dandelion (*Taraxacum officinale*). Natural Standard
Database Web site. Accessed at
www.naturalstandard.com on June 9, 2009.

Dandelion root with herb. In: Blumenthal M,
Goldberg A, Brinckman J, eds. *Herbal
Medicine: Expanded Commission E*

Monographs. Newton, MA: Lippincott Williams & Wilkins; 2000:359–366.

Dandelion. Natural Medicines Comprehensive Database Web site. Accessed at www.naturaldatabase.com on June 9, 2009.

Echinacea (*E. angustifolia DC, E. pallida, E. purpurea*). Natural Standard Database Web site. Accessed at www.naturalstandard.com on June 1, 2009.

Echinacea. In: Blumenthal M, Goldberg A, Brinckman J, eds. *Herbal Medicine: Expanded Commission E Monographs*. Newton, MA: Lippincott Williams & Wilkins; 2000:88–102.

Echinacea. Natural Medicines Comprehensive Database Web site. Accessed at www.naturaldatabase.com on May 11, 2009.

Evening primrose oil (*Oenothera biennis* L.). Natural Standard Web site. Accessed at www.naturalstandard.com on June 11, 2009.

Evening primrose oil. Natural Medicines Comprehensive Database Web site. Accessed at www.naturaldatabase.com on June 11, 2009.

Foster S, Duke JA. *A Field Guide toMedicinal Plants and Herbs of the Eastern and Central US*. Boston, Mass: Houghton Mifflin; 2000:321-323.

Fugh-Berman A. Echinacea for the prevention and treatment of upper respiratory infections. *Seminars in Inative Medicine*. 2003;1(2):1tegr06–111.

Fugh-Berman A. Herb-drug interactions. *Lancet*. 2000; 355(9198):134-138.

Hardy ML. Herbs of special interest to women. *J Am Pharm Assoc*. 2000;40(2):234-242.

Hoffmann D. *Therapeutic Herbalism*. Santa Cruz, Calif: Therapeutic Herbalism Press; 2000.

Kuhn MA, Winston D. *Herbal Therapy and Supplements*. Philadelphia, Pa: Lippincott; 2001.

Mahady GB, Low Dog T, Barrett ML, et al. United States Pharmacopeia review of the black cohosh case reports of hepatotoxicity. *Menopause*. 2008;15(4 Pt 1):628–638.

McGuffin M, Hobbs C, Upton R, et al, eds. *American Herbal Products Association's Botanical*

Safety Handbook. Boca Raton, Fla: CRC
Press; 1997:101.

Milk thistle (*Silybum marianum*), silymarin. Natural
Standard Database Web site. Accessed at
www.naturalstandard.com on October 7,
2009.

Milk thistle (*Silybum marianum*). In: Coates P,
Blackman M, Cragg G, et al., eds.
Encyclopedia of Dietary Supplements. New
York, NY: Marcel Dekker; 2005:467–482.

Milk thistle fruit. In: Blumenthal M, Goldberg A,
Brinckman J, eds. *Herbal Medicine:
Expanded Commission E Monographs*.
Newton, MA: Lippincott Williams & Wilkins;
2000:257–263.

Milk thistle. Natural Medicines Comprehensive
Database. Accessed at
www.naturaldatabase.com on October 7,
2009.

National Center for Complementary and Alternative
Medicine and Office of Dietary
Supplements. *Workshop on the Safety of
Black Cohosh in Clinical Studies*. National
Center for Complementary and Alternative
Medicine Web site. Accessed at

nccam.nih.gov/news/events/blackcohosh/blackcohosh_mtngsumm.htm on June 3, 2010.

National Center for Complementary and Alternative Medicine, National Institute of Health

Newton KM, Reed SD, LaCroix AZ, et al. Treatment of vasomotor symptoms of menopause with black cohosh, multibotanicals, soy, hormone therapy, or placebo: a randomized trial. *Annals of Internal Medicine*. 2006;145(12):869–879.

Office of Dietary Supplements. *Dietary Supplement Fact Sheet: Black Cohosh*. Office of Dietary Supplements Web site. Accessed at ods.od.nih.gov/factsheets/blackcohosh.asp on June 3, 2010.

Red clover (*Trifolium pratense*). In: Coates P, Blackman M, Cragg G, et al., eds. *Encyclopedia of Dietary Supplements*. New York, NY: Marcel Dekker; 2005:587–602.

Red clover (*Trifolium pratense*). Natural Standard Database Web site. Accessed at www.naturalstandard.com on July 22, 2009.

Red clover. Natural Medicines Comprehensive
Database Web site. Accessed at
www.naturaldatabase.com on July 22,
2009.

Saw palmetto (*Serenoa repens* [Bartran] Small).
Natural Standard Database Web site.
Accessed at www.naturalstandard.com on
August 4, 2009.

Saw palmetto (*Serenoa repens*). In: Coates P,
Blackman M, Cragg G, et al., eds.
Encyclopedia of Dietary Supplements. New
York, NY: Marcel Dekker; 2005;635–644.

Saw palmetto. Natural Medicines Comprehensive
Database Web site. Accessed at
www.naturaldatabase.com on August 7,
2009.

Shahidi F, Miraliakbari H. Evening primrose
(*Oenothera biennis*). In: Coates P, Blackman
M, Cragg G, et al., eds. *Encyclopedia of
Dietary Supplements*. New York, NY: Marcel
Dekker; 2005:197–210.

Saw palmetto berry. In: Blumenthal M, Goldberg A,
Brinckman J, eds. *Herbal Medicine:
Expanded Commission E Monographs*.
Newton, MA: Lippincott Williams & Wilkins;
2000:335–340.

Printed in Great Britain
by Amazon.co.uk, Ltd.,
Marston Gate.